the four elements

air

María Rius
J. M. Parramón

BARRON'S

New York • Toronto • Sydney

You don't see it, but it moves.

It's called a *breeze* when it caresses.

a *wind* when it bothers you,

and a *hurricane* when it destroys.

It has the power to raise a kite...

...to make a balloon fly...

...to move the sails of a windmill...

...to push a boat...

...to hold up a plane...

...to let birds fly...

...to let a man come down from the sky...

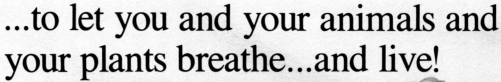

...to let you and your animals and your plants breathe...and live!

IT'S AIR!

AIR

"A I R
A I - R
A - I R
A - I - R"

The air that surrounds us

You can't see it, you can't touch it, but it's there. It's what shakes the branches of the trees, lifts up skirts, ruffles your hair.

It's what you put in a football or a bicycle tire. Air can be compressed. The ball regains its shape after a kick. Air is elastic. When a car tire is punctured, the air escapes, it spreads. How far?

Does air end?

Imagine that air is formed by tiny balls that move in every direction without stopping, bumping into each other and against the wall of the container that holds them. Air surrounds us.

But air doesn't scatter through interplanetary space because the force of gravity pulls on each of the little balls. Air has weight. The thickness of the layer of air that surrounds the earth (the atmosphere) is about seven hundred miles.

The weight of the air at the earth's surface is called *atmospheric pressure*. Where is this pressure the greatest? At the top of a mountain or at sea level? As we go up, the air becomes less dense and the atmospheric pressure decreases, so the support layer of air is thinner. On a very high mountain you will find it difficult to breathe.

The air we breathe

You couldn't stand being underwater for a long time: you need air to breathe. You aren't like fish who can absorb the oxygen dissolved in the water.

The air is composed of about one-fifth oxygen and four-fifths nitrogen, plus some other substances.

Oxygen is indispensable in all combustion. Therefore, if you have a lit candle and you cover it with a glass container, you'll see that after a little while it will go out. Most of the oxygen has been consumed!

In the olden days they called nitrogen "azoe" from the Greek *a* (negation) and *zoe* (life), because with only nitrogen the candle would go out and the animals would die. But its presence is important biologically. Some microscopic bacteria, living in the soil, are capable of using nitrogen from the air to make nitrate compounds, which are necessary for plant development.

When you take a cold bottle from the refrigerator, little drops appear on the outside of the bottles. What are they? These little drops are water that are found in the air in the form of vapor, which has condensed upon contact with a cold surface.

In small and varying proportions, there are other gases in the air, such as water vapor and carbon dioxide, the gas produced by your body in respiration, and products formed by many combustions, such as the gas we cook with. In the presence of light, green plants are capable of transforming substances found in the soil into food for themselves and also of renewing the oxygen in the atmosphere.

Furthermore, air contains in suspension the dust which can be seen when a ray of sunshine enters a dark room. This dust is the result of wind erosion.

Why does the wind blow?

If you put a pinwheel (like the ones sold at fairs) on top of a radiator or a lit stove, you'll see how its blades spin. Warm air is lighter than cold air; it rises and is replaced by the descending cold air. You may have seen a warm air balloon rise into the sky.

The same happens — on a larger scale — in the atmosphere. The sun does not warm all the parts of the earth evenly, causing displacement of the air that surrounds it. Wind is air in movement. The breezes that originate in the coastal areas are the result of the temperature difference between the air mass over the sea and the mass over the land.

The energy that the air possesses, owing to its movement, is known by the name *kinetic* energy. It is a cheap and renewable form of energy, used in the olden days to move ships, windmills, etc. It was replaced by other forms of energy, but now there are intentions to revive it, most of all in areas where wind is strong.

Pure air

How many times, on the top of a mountain, have we exclaimed "What pure air!" and how many other times has the gas from an exhaust pipe or cigarette smoke made us cough.

In the city, green areas are diminishing because of so much land development. These green areas have the role, among others, of renewing the polluted air.

Air is the fluid in which all we living beings are immersed. The cleaner we keep it, the fewer illnesses we will suffer and the happier will be our existence.